Japan's Satellite Program Book

A Comprehensive Journey from JAXA's H3 Rockets to SpaceX's Falcon 9 and Beyond

By

Kai Astral

Table of contents.

1.0

Introduction

Historical Context:

Japan's Early Challenges in Space Exploration

In tracing the historical context of Japan's space exploration endeavors, it is imperative to delve into the nation's early, narrative that unfolds against the backdrop of post-war recovery and technological resurgence. Emerging from the devastation of World War II, Japan faced profound socio-economic struggles, with the immediate aftermath marked by the daunting task of reconstruction and redefinition on multiple fronts.

The early post-war era witnessed a Japan seeking to rebuild its identity, not only as an economic powerhouse but also as a player in the technological frontier. Amidst this intricate spectrum of recovery, the ambition to venture into space exploration emerged as a symbol of Japan's resilience and determination to embrace innovation. However, the journey was fraught with challenges, both internal and external.

Internally, Japan grappled with the aftermath of wartime restrictions and the need to redirect its scientific and technological capabilities towards peaceful pursuits. The Japanese scientific community, once constrained by wartime policies, now found itself at the crossroads of a transformative era, seeking ways to harness its intellectual capital for the greater good.

Externally, the geopolitical landscape presented formidable challenges. The Cold War era defined the space race, with the United States and the Soviet Union dominating the cosmos. Japan, in the shadow of these superpowers, navigated a delicate balance between forging its space ambitions and aligning with global dynamics. The absence of a space program during the early years underscored the intricacies of Japan's position on the international stage.

Furthermore, Japan's approach to space exploration was not just a scientific endeavor; it mirrored the nation's broader aspirations for global recognition and technological prowess. The 1950s and 1960s saw Japan taking its initial steps with sounding rockets, a modest beginning that laid the groundwork for future endeavors. However, these early efforts were not without setbacks, as Japan encountered technological hurdles and faced skepticism about its capabilities.

The transformational turning point came with the establishment of the National Space Development Agency of Japan (NASDA) in 1969, a pivotal moment that signaled Japan's commitment to venturing into the cosmos. The agency's inception marked the consolidation of national efforts, bringing together scientific minds, engineers, and policymakers under a unified banner. This marked a departure from the fragmented early attempts, symbolizing Japan's resolve to overcome challenges through coordinated endeavors.

As Japan embarked on this journey, the historical context shaped the narrative of its space exploration. The early challenges, while formidable, laid the foundation for resilience, innovation, and the eventual triumphs that would unfold in the years to come. This historical lens provides a nuanced understanding of the intricate spectrum that defines Japan's trajectory in space exploration, setting the stage for the comprehensive exploration of its satellite program from defeat to triumph.

Transformation and Triumph:
Setting the Stage for the Satellite Program

The narrative of Japan's space exploration takes a profound turn as we delve into the era of transformation and triumph, a pivotal juncture that laid the very foundations for the nation's ambitious satellite program, epitomized by the remarkable development of JAXA's H3 Rockets. This epoch, marked by technological ascendancy and strategic vision, unravels a compelling saga of how Japan metamorphosed from the challenges of its early forays into space into a dynamic force in the global space arena.

Amidst the echoes of post-war recovery, Japan underwent a metamorphic evolution in the latter half of the 20th century. The establishment of the National Space Development Agency of Japan (NASDA) in 1969 was a watershed moment that heralded a new era of focus, collaboration, and strategic planning in the realm of space exploration. This agency, which later became part of the Japan Aerospace Exploration Agency (JAXA) in 2003, became the crucible in which Japan's space aspirations were forged.

The transformation was not merely institutional; it was a comprehensive reorientation of Japan's scientific and technological landscape. A confluence of factors, including geopolitical shifts, economic resurgence, and a burgeoning national ambition, converged to catalyze Japan's propulsion into the forefront of space innovation. The nation transitioned from tentative steps with sounding rockets to envisioning and crafting its indigenously developed satellites.

The triumphs of this transformative era find resonance in the ambitious development of JAXA's H3 Rockets. These rockets, a testament to Japan's technical prowess and commitment to space exploration, symbolize the culmination of decades of perseverance and innovation. The H3 Rockets are not just vehicles designed for propelling payloads into space; they are the tangible expressions of Japan's resilience, precision engineering, and strategic acumen.

The intricacies of the H3 Rockets unfold against the backdrop of collaborative efforts, both domestic and international. Japan's engagement in joint ventures, such as the LUPEX project with India and contributions to the U.S.-led Artemis moon exploration program, underscores the nation's commitment to global cooperation in the pursuit of scientific discovery and exploration beyond terrestrial boundaries.

The development of H3 Rockets represents not only a technological milestone but also a strategic imperative for Japan. As satellite launch demands surge globally, these rockets position Japan as a key player in the commercial space industry. The vision extends beyond national boundaries, shaping Japan's role in the broader narrative of space exploration, where the

H3 Rockets become emissaries of innovation and symbols of triumphant scientific achievement.

In essence, the transformation and triumph encapsulated in this period set the stage for Japan's satellite program and the evolution of the H3 Rockets. This introduction serves as a prelude to a comprehensive exploration of the technological intricacies, collaborative ventures, and the broader implications of Japan's journey from the challenges of its early space endeavors to the triumphant era of cutting-edge satellite launch capabilities.

2.0

JAXA's H3 Rockets

Unveiling Japan's Technological

Advancements

The narrative of Japan's journey into the cosmos finds its most vivid expression in the unveiling of the nation's technological advancements, exemplified by the remarkable development of JAXA's H3 Rockets. This technological odyssey represents a convergence of scientific ingenuity, engineering precision, and a national commitment to establishing Japan as a formidable player in the competitive space exploration arena.

At the core of Japan's technological advancements lies a spectrum woven with threads of innovation and persistence. The country's ascent from early challenges to pioneering the H3 Rockets is a testament to the collaborative spirit that permeates Japan's scientific and engineering communities. It's a story of researchers, engineers, and visionaries who tirelessly worked to push the boundaries of what was deemed possible.

The H3 Rockets, a crowning achievement in this narrative, embody a spectrum of cutting-edge technologies. From propulsion systems to payload deployment mechanisms, every facet of these rockets reflects Japan's unwavering commitment to pushing the envelope of space exploration capabilities. The utilization of advanced materials, precision manufacturing, and state-of-the-art navigation systems position the H3 Rockets at the forefront of technological innovation.

A critical aspect of this technological unveiling is the adaptability and versatility of the H3 Rockets. Designed to accommodate various payloads and mission requirements, these rockets showcase Japan's strategic foresight in catering to the diverse needs of the evolving space industry. Whether launching satellites for domestic purposes or contributing to international collaborative projects, the H3 Rockets signify a modular and flexible approach to space exploration.

The technological narrative extends beyond the physical components of the rockets; it encapsulates the sophisticated ground control systems and data analytics that orchestrate each mission. Japan's prowess in data processing and communication technologies ensures that the H3 Rockets not only reach their intended destinations but also transmit invaluable scientific data back to Earth, contributing to the collective understanding of the cosmos.

Moreover, the unveiling of Japan's technological advancements through the H3 Rockets is not merely a national achievement but a global milestone. As these rockets become integral contributors to collaborative ventures like the LUPEX project with India and the U.S.-led Artemis moon exploration program, Japan positions itself as a reliable partner on the international stage, fostering a spirit of cooperation in the pursuit of scientific knowledge beyond borders.

In essence, the technological unveiling represented by JAXA's H3 Rockets is a nuanced symphony of innovation, precision, and strategic vision. It signifies not only Japan's technological capabilities but also its commitment to playing a pivotal role in shaping the future of space exploration. This chapter in Japan's space narrative marks a culmination of technological achievements that reverberate far beyond the confines of Earth's atmosphere, resonating as a beacon of human ingenuity reaching for the stars.

Collaborative Endeavors:
LUPEX Project and Artemis Contributions

The intricate dance of international collaboration takes center stage in the narrative of JAXA's H3 Rockets, as Japan, through its technological prowess, actively engages in two noteworthy collaborative endeavors: the LUPEX Project with India and contributions to the U.S.-led Artemis moon exploration program. These collaborative ventures not only underscore Japan's commitment to global cooperation in space exploration but also position the H3 Rockets as instrumental components in these ambitious international initiatives.

The LUPEX Project, a collaborative effort between Japan and India, represents a paradigm of cooperative space exploration. This venture aims to send a lunar explorer into space, conducting scientific investigations on the Moon's surface. Japan's participation in the LUPEX Project exemplifies the nation's willingness to pool resources, knowledge, and expertise with international partners, fostering a shared commitment to unraveling the mysteries of the lunar landscape.

Contributions to the U.S.-led Artemis moon exploration program further highlight Japan's integral role in shaping the future of human space exploration. As a key participant, Japan leverages the capabilities of its H3 Rockets to transport cargo spacecraft essential for the Artemis program. This collaborative effort not only showcases Japan's technological contributions but also solidifies its position as a trusted partner in advancing the frontiers of human space exploration.

The significance of these collaborative endeavors extends beyond the mere transport of payloads. They symbolize a broader ethos of shared goals and mutual understanding in the pursuit of scientific discovery. Adding Japan's H3 Rockets to these international projects makes the rockets even more useful. They are now carrying more than just cargo; they are also carrying cooperation, diplomacy, and humanity's quest to explore and understand the universe.

Moreover, the collaborative nature of these projects fosters a synergy of diverse perspectives and expertise. Japan's contributions to the LUPEX Project and Artemis program exemplify the idea that space exploration is a collective endeavor, transcending geopolitical boundaries. By actively participating in these initiatives, Japan not only furthers its scientific objectives but also contributes to the broader spectrum of human exploration, where nations unite in the pursuit of knowledge beyond Earth.

The collaborative endeavors surrounding JAXA's H3 Rockets, be it the LUPEX Project with India or contributions to the Artemis program, paint a picture of a global community striving for a shared vision of space exploration. The H3 Rockets, as key players in these ventures, become ambassadors of cooperation, forging pathways to a future where the mysteries of space are unraveled through collective effort, and the boundaries of exploration extend far beyond the limits of any single nation.

3.0

The Spacecraft Revolution

Overview of Modern Satellite Launch Vehicles

The landscape of modern satellite launch vehicles heralds a revolution in space exploration, reflecting an amalgamation of technological innovation, strategic foresight, and a burgeoning demand for reliable and efficient access to space. This paradigm shift encapsulates a comprehensive overview of the evolution of satellite launch vehicles, portraying a journey from early space endeavors to the cutting-edge spacecraft revolution that characterizes the contemporary era.

The evolution of satellite launch vehicles mirrors the ever-expanding ambitions of humankind to explore and harness the potential of outer space. From the rudimentary beginnings of single-use rockets, the space industry has undergone a transformative journey marked by incremental advancements, culminating in the development of highly sophisticated and reusable launch vehicles.

The introduction of the Space Shuttle by NASA in the 1980s exemplified a pioneering leap in space transportation technology. This reusable spacecraft not only demonstrated the feasibility of launching payloads into space multiple times but also laid the groundwork for the subsequent evolution of modern satellite launch vehicles. The concept of reusability, a cornerstone of the spacecraft revolution, has since become a focal point for major players in the global space industry.

The advent of private companies entering the space race has injected a new dimension of competition and innovation. SpaceX, with its Falcon 9 rocket, stands as a trailblazer in this new era. The Falcon 9's reusability, exemplified by its ability to return to Earth and be relaunched, has not only significantly reduced launch costs but has also set a precedent for sustainable space exploration.

The spacecraft revolution is further propelled by advancements in propulsion systems, materials science, and miniaturization of satellite technology. Miniaturized satellites, colloquially known as CubeSats, have democratized access to space, allowing smaller organizations and even educational institutions to participate in satellite-based research and experiments. This democratization, in turn, fuels the demand for more versatile and cost-effective launch vehicles.

In this dynamic landscape, JAXA's H3 Rockets play a crucial role, representing Japan's commitment to staying at the forefront of space exploration. The H3 Rockets bring to the table a blend of reliability, efficiency, and adaptability, contributing to the diversity of options available for launching satellites into orbit.

As we navigate the overview of modern satellite launch vehicles, it becomes evident that the spacecraft revolution is not merely a technological evolution; it signifies a paradigm shift in how we perceive and interact with space.

The emphasis on reusability, cost-effectiveness, and accessibility opens up new possibilities for scientific exploration, commercial ventures, and international collaboration.

In essence, the spacecraft revolution encapsulates a transformative chapter in the history of space exploration, where traditional boundaries are pushed, and the cosmos becomes an arena accessible to a broader spectrum of participants. The ongoing developments in modern satellite launch vehicles, propelled by technological innovations and a changing economic landscape, lay the groundwork for a future where space becomes not just a frontier for a select few but a realm open to the collective aspirations of humanity.

Highlighting the Significance of JAXA's Contribution

In the grand spectrum of the spacecraft revolution, JAXA's contributions emerge as a significant thread, weaving through the fabric of innovation, collaboration, and progress in the realm of space exploration. As we delve into the intricate details, it becomes apparent that the significance of JAXA's role extends far beyond the borders of Japan; it resonates globally, shaping the trajectory of the spacecraft revolution and influencing the broader narrative of human engagement with outer space.

At the heart of JAXA's contributions lies a commitment to advancing space exploration capabilities. The development of the H3 Rockets exemplifies Japan's dedication to staying at the forefront of technological innovation in launching payloads into space. The H3 Rockets, with their cutting-edge design and modular features, underscore the nation's emphasis on adaptability, reliability, and efficiency, all of which are paramount in the evolving landscape of the spacecraft revolution.

JAXA's role in the spacecraft revolution becomes particularly pronounced when considering the global collaboration fostered by projects like the LUPEX initiative with India and contributions to the U.S.-led Artemis moon exploration program. By actively engaging in these international endeavors, Japan not only showcases the capabilities of its space launch vehicles but also fosters a spirit of cooperation, uniting nations in the pursuit of shared scientific objectives.

Furthermore, JAXA's contributions extend to addressing the demands of an increasingly competitive and commercialized space industry. The spacecraft revolution is characterized not only by scientific exploration but also by the burgeoning market for satellite launches. In this context, the H3 Rockets position Japan as a key player, offering reliable and cost-effective solutions that cater to the diverse needs of the commercial space sector.

The significance of JAXA's contributions is also underscored by the broader implications for space policy and diplomacy. As nations assert their presence in outer space, JAXA's endeavors contribute to shaping a narrative where space becomes a domain for peaceful collaboration and shared discovery. Japan's active participation in international projects and collaborations serves as a testament to its commitment to leveraging space exploration as a means of fostering goodwill and understanding among nations.

Moreover, JAXA's role in the spacecraft revolution aligns with the global trend towards sustainability and environmental consciousness. The H3 Rockets, designed with a focus on reusability and efficiency, echo the broader shift in the space industry towards environmentally responsible practices. By prioritizing these principles, JAXA contributes not only to technological advancement but also to the responsible stewardship of Earth's resources.

In essence, JAXA's contributions to the spacecraft revolution encapsulate a multifaceted narrative. They represent a technological triumph, a diplomatic overture, a commercial endeavor, and a commitment to sustainability. The significance of Japan's role in shaping the trajectory of the spacecraft revolution is a testament to the nation's vision, resilience, and dedication to the noble pursuit of exploring the cosmos for the betterment of humanity.

4.0

Global Competitors and Collaborators

ULA's Vulcan

A Joint Venture by Boeing and Lockheed Martin, in the dynamic landscape of global space exploration, ULA's Vulcan rocket stands as a testament to collaborative innovation, representing a joint venture between aerospace giants Boeing and Lockheed Martin. This partnership not only underscores the technological prowess of the United States but also exemplifies the strategic collaboration necessary for navigating the complexities of the modern space industry.

The genesis of ULA's Vulcan rocket can be traced back to the evolving demands of the space sector. In a landscape where competition is fierce and technological advancements are rapid, collaboration becomes a key driver for success. Boeing and Lockheed Martin, both titans in the aerospace industry, recognized the need to pool their resources, expertise, and capabilities to create a launch vehicle that could not only meet the contemporary demands of satellite launches but also pave the way for future exploration endeavors.

The Vulcan rocket, as a result of this collaboration, represents a synthesis of the best that Boeing and Lockheed Martin bring to the table. Its design incorporates advanced technologies, efficient manufacturing processes, and a modular architecture that allows for adaptability to a range of mission requirements. The rocket's versatility positions it as a formidable competitor in the global market for satellite launches, where adaptability and reliability are paramount.

Moreover, ULA's Vulcan is a strategic response to the changing dynamics of the space industry, particularly in the face of competition from emerging players. As new entrants like SpaceX revolutionize space access with reusable rocket technology, ULA's joint venture ensures that the United States remains at the forefront of innovation, offering a competitive alternative to meet the diverse needs of both government and commercial satellite launches.

The collaborative nature of ULA's Vulcan also echoes a broader trend in the space industry – the recognition that no single entity can navigate the complexities of space exploration in isolation. The challenges of developing and operating reliable launch vehicles, exploring new frontiers, and sustaining a competitive edge necessitate partnerships that leverage the strengths of multiple entities. In this context, ULA's joint venture becomes not just a technological achievement but a strategic imperative for staying ahead in the global space race.

The global landscape of satellite launches is characterized by a multitude of players, each vying for a share of the market. ULA's Vulcan, through the collaborative efforts of Boeing and Lockheed Martin, emerges as a key player in this arena. Its contributions extend beyond national boundaries, as the United Launch Alliance actively engages in commercial contracts, government missions, and international partnerships, contributing to the broader narrative of global collaboration in space exploration.

As we examine ULA's Vulcan in the context of global competitors and collaborators, it becomes evident that the joint venture between Boeing and Lockheed Martin represents a harmonious blend of expertise, innovation, and strategic vision. In the intricate dance of global space exploration, where the spacecraft revolution is reshaping traditional paradigms, ULA's Vulcan stands as a symbol of the power of collaboration to propel humanity's endeavors beyond Earth's atmosphere.

Ariane 6 from the European Space Agency:

International Collaboration in Space Exploration

In the expansive panorama of global space exploration, the Ariane 6 rocket, developed under the auspices of the European Space Agency (ESA), serves as a stellar example of international collaboration, highlighting the cooperative efforts of multiple European countries in advancing space access capabilities. The development and deployment of the Ariane 6 represent not only a technological milestone but also underscore the significance of collaborative initiatives in navigating the challenges and opportunities of the modern space industry.

The genesis of the Ariane 6 can be traced back to the recognition that space exploration is an endeavor that transcends national boundaries. The European Space Agency, an intergovernmental organization comprising 22 member states, operates on the principle that pooling resources, knowledge, and technological capabilities is not only efficient but also essential for maintaining a competitive edge in space exploration. The Ariane 6 project exemplifies this philosophy, drawing on the expertise and contributions of multiple European nations to create a launch vehicle that meets the demands of a rapidly evolving space market.

The collaborative nature of the Ariane 6 project extends beyond the development phase and encompasses the entire lifecycle of the rocket.

From manufacturing to launch operations, various European countries actively contribute to different aspects of the project, fostering a sense of shared responsibility and mutual benefit. This distributed approach not only leverages the specific strengths of each participating nation but also ensures a robust and resilient space infrastructure.

Furthermore, the Ariane 6 serves as a strategic response to the intensifying competition in the global space market. With emerging players and increased commercialization, Europe recognizes the need for a launch vehicle that is not only technologically advanced but also economically viable. The collaboration with the European Space Agency allows for a cost-effective approach to space exploration, ensuring that the

Ariane 6 can offer competitive solutions for both government and commercial satellite launches.

The significance of the Ariane 6 in the global context lies in its technical specifications and its role as a symbol of European unity in space exploration. The rocket represents a joint venture where nations set aside individual interests to work towards a collective goal. In an era where geopolitical tensions can influence international collaborations, the success of the Ariane 6 demonstrates the resilience and effectiveness of collaborative efforts in overcoming challenges and achieving shared objectives.

As the European Space Agency actively engages in commercial contracts, international partnerships, and government missions, the Ariane 6 becomes a key player in the broader narrative of global competition and collaboration in space exploration. Its contributions extend beyond Europe, influencing the dynamics of the entire space industry and showcasing the importance of shared endeavors in advancing our understanding of the cosmos.

The Ariane 6 from the European Space Agency stands as a testament to the power of international collaboration in space exploration. Its development and deployment exemplify the collective aspirations of European nations to navigate the complexities of the modern space industry, offering a competitive and collaborative alternative in the global arena of satellite launches and shaping the narrative of humanity's continued journey into the cosmos.

5.0

Satellite Launch Demand Surge

Analyzing Factors Driving Global Demand

The surge in satellite launch demand on a global scale is a multifaceted phenomenon, driven by a confluence of factors that collectively shape the trajectory of the space industry. This surge represents a pivotal moment in the evolution of space exploration, reflecting not only the expanding role of satellites in various sectors but also the transformative impact of technological advancements and shifting paradigms in the utilization of space assets.

At the forefront of the factors propelling this surge is the proliferation of satellite applications across diverse domains. Satellites have evolved from being primarily tools of communication and navigation to becoming indispensable assets in fields such as Earth observation, weather monitoring, scientific research, and national security. The increasing reliance on satellite technology for real-time data, global connectivity, and comprehensive Earth observation has fueled a heightened demand for launching satellites into orbit.

The growing appetite for connectivity in our interconnected world is a significant driver of satellite launch demand. With the rise of the Internet of Things (IoT), 5G technology, and the global demand for high-speed broadband, satellites play a pivotal role in ensuring ubiquitous and reliable connectivity, especially in remote or underserved regions. The deployment of satellite constellations by private companies aiming to provide global broadband coverage further amplifies the demand for frequent and cost-effective satellite launches.

Moreover, the surge in satellite launch demand is intrinsically linked to the commercialization of space. The emergence of private space companies, such as SpaceX, Blue Origin, and others, has disrupted traditional models, introducing competition, reducing launch costs, and increasing the frequency of launches. The advent of reusable rocket technology, exemplified by SpaceX's Falcon 9, has particularly played a pivotal role in driving down launch costs and making space more accessible for a broader range of missions.

The democratization of space access is another critical factor. As more countries enter the arena of space exploration, driven by scientific, economic, and strategic imperatives, the demand for launching satellites for national purposes has surged. Nations are increasingly leveraging space assets for applications ranging from communications and Earth observation to scientific research and national security, contributing to a diverse and dynamic global demand landscape.

Furthermore, the surge in satellite launch demand is closely tied to the aspirations of emerging space nations and the rise of new players in the space industry. Countries that traditionally had limited space programs are now investing in space capabilities, seeking to harness the benefits of satellite technology for economic development, disaster management, and scientific research. This trend not only reflects the growing recognition of the strategic importance of space but also contributes significantly to the increasing demand for satellite launches.

The space industry's response to environmental concerns and sustainability is another facet of the surge in satellite launch demand. With a growing emphasis on responsible space practices, there is an increased focus on developing small satellites and CubeSats, which are more cost-effective and have a reduced environmental impact. This shift towards miniaturized satellites is reshaping the demand landscape, requiring more frequent launches to deploy these smaller payloads into orbit.

In conclusion, the surge in satellite launch demand is a complex and dynamic phenomenon shaped by a myriad of interconnected factors. From the evolving role of satellites in various sectors to the commercialization of space, the democratization of space access, and the rise of emerging space nations, these factors collectively contribute to a landscape where the demand for satellite launches is not only significant but also reflects the transformative nature of the space industry in the 21st century.

Japan's Response to the Rise of Affordable Commercial Vehicles

Japan's response to the rise of affordable commercial vehicles in the context of the satellite launch demand surge reflects a strategic recalibration that intertwines technological innovation, economic considerations, and the imperative to maintain a competitive edge in the evolving space industry. As the dynamics of space access undergo a transformative shift driven by private companies offering cost-effective launch solutions, Japan, through organizations like the Japan Aerospace Exploration Agency (JAXA) and private enterprises, is adapting its approach to satellite launches to meet the challenges and opportunities presented by this paradigm shift.

At the heart of Japan's response lies a recognition of the changing landscape in the space launch sector, where the emergence of private entities like SpaceX has disrupted traditional models. The cost efficiencies achieved through reusable rocket technology, pioneered by companies such as SpaceX with the Falcon 9, have led to a reduction in launch costs and an increase in launch frequency. In response to this trend, Japan is strategically positioning itself to leverage both its traditional strengths and innovative capabilities to remain competitive in the global satellite launch market.

One key aspect of Japan's response is the acknowledgment of the importance of cost-effectiveness and efficiency in satellite launches. As affordable commercial vehicles gain prominence, Japan is investing in research and development to enhance the cost efficiency of its own launch vehicles. Initiatives to explore reusable rocket technology, streamlined manufacturing processes, and overall operational efficiency are integral components of Japan's strategy to stay competitive in the face of changing market dynamics.

Furthermore, Japan is fostering collaboration between public and private entities to amplify its response to the surge in satellite launch demand. The synergy between government agencies like JAXA and private companies ensures a comprehensive approach that combines the technological expertise of JAXA with the agility and market-driven focus of private enterprises. This collaborative effort not only enhances the nation's ability to meet the demand for satellite launches but also fosters innovation in the rapidly evolving space sector.

Another dimension of Japan's response involves the diversification of its launch capabilities. Recognizing the demand for launching smaller payloads, Japan is actively exploring the development of small satellite launch vehicles that cater to the evolving needs of the space industry. By focusing on miniaturized satellite technology, Japan aligns its launch capabilities with the trends favoring smaller, more cost-effective satellites and contributes to the ongoing democratization of space access.

Additionally, Japan's response is characterized by a proactive engagement in international collaborations. The nation actively participates in joint ventures, such as the aforementioned LUPEX project with India and contributions to the U.S.-led Artemis moon exploration program. These collaborations not only contribute to the global pursuit of scientific knowledge but also position Japan as a reliable partner in the broader narrative of space exploration.

Moreover, Japan is emphasizing the commercialization of its space capabilities. By actively pursuing commercial contracts and partnerships with private entities, Japan seeks to leverage its technological expertise to provide launch services for a diverse range of clients. This commercial approach not only contributes to the nation's economic growth but also ensures that Japan remains an active player in the increasingly commercialized space industry.

In conclusion, Japan's response to the rise of affordable commercial vehicles in the context of the satellite launch demand surge is a nuanced and multifaceted strategy. It involves a careful balance of technological innovation, collaboration, and commercialization to navigate the changing dynamics of the global space industry. By adapting to the challenges and opportunities presented by affordable commercial launch solutions, Japan endeavors to secure its position as a key player in the competitive and dynamic landscape of satellite launches.

6.0

Economic Impact and International Collaboration

Examining Financial Growth in Japan's Space Industry

Examining the financial growth in Japan's space industry unveils a multifaceted narrative that intertwines economic impact and international collaboration. As Japan asserts itself as a formidable player in the global space arena, the growth trajectory of its space industry reflects not only technological advancements and strategic initiatives but also the economic ripple effects and diplomatic significance of a flourishing space sector.

At the core of this examination lies the economic impact of Japan's investments in space exploration and technology. The financial growth in the country's space industry is characterized by substantial government funding, private investments, and a burgeoning commercial space sector. The Japanese government, through entities like the Japan Aerospace Exploration Agency (JAXA), has consistently allocated substantial budgets to space-related initiatives. This financial commitment has not only fueled research and development but has also stimulated economic activity across various sectors, from manufacturing to technology.

The economic impact extends beyond government initiatives to private enterprises within Japan's space industry. Companies engaged in satellite manufacturing, launch services, and satellite applications contribute significantly to the country's economic growth. The influx of private investments into space startups and established companies has created a dynamic ecosystem that fosters innovation, job creation, and the development of cutting-edge technologies. The economic landscape of Japan's space industry, therefore, mirrors a broader trend where space becomes not just a scientific frontier but also an economic driver.

Moreover, the financial growth in Japan's space industry is intricately linked to the expansion of commercial space activities. The rise of private companies engaging in satellite launches, space tourism, and satellite-based services adds a commercial dimension to the industry. The economic potential of commercial space endeavors is substantial, creating new revenue streams, generating employment opportunities, and positioning Japan as a hub for space-related business activities.

International collaboration plays a pivotal role in amplifying the economic impact of Japan's space industry. Engaging in joint ventures, collaborative projects, and partnerships with other spacefaring nations contributes not only to the sharing of scientific knowledge and technological expertise but also to economic collaboration. The financial growth is not confined to national borders; it is augmented by the economic synergies created through collaborative initiatives, such as joint satellite development programs and participation in international space exploration projects.

The economic impact of Japan's space industry is further accentuated by its role as a reliable partner in the global space market. Japan's capabilities in satellite manufacturing, launch services, and space exploration make it an attractive collaborator for countries and organizations seeking to leverage its technological expertise. This collaboration not only enhances Japan's economic standing but also contributes to the nation's soft power and diplomatic influence on the international stage.

Furthermore, the economic growth in Japan's space industry is reflective of a broader trend where space is increasingly perceived as an economic frontier. As satellite applications become integral to various industries, from agriculture and environmental monitoring to telecommunications and navigation, the economic significance of space-related activities becomes more pronounced. Japan's strategic positioning in this evolving landscape ensures that it remains at the forefront of economic opportunities emerging from the commercialization of space.

In conclusion, examining the financial growth in Japan's space industry unveils a narrative that transcends mere economic statistics. It is a story of strategic investments, private entrepreneurship, international collaboration, and the economic ripple effects of a flourishing space sector. As Japan continues to navigate the economic dynamics of space exploration, the synergy between technological innovation, economic impact, and global collaboration positions the country as a key player in the ongoing evolution of the global space industry.

Government Initiatives, Regulations, and Collaborative Efforts on the Global Stage

The intertwining realms of government initiatives, regulations, and collaborative efforts on the global stage significantly shape the economic impact and international collaboration within the evolving landscape of space exploration. As nations recognize the strategic and economic importance of space activities, governments play a pivotal role in orchestrating initiatives, formulating regulations, and fostering collaborative endeavors that not only stimulate economic growth but also define the trajectory of international cooperation in space.

Government initiatives in the realm of space exploration are multifaceted, ranging from scientific missions and technological development to economic incentives and strategic investments. Governments across the globe, from established space powers to emerging players, allocate significant budgets to space agencies or entities dedicated to advancing space exploration objectives. These initiatives serve as catalysts for technological innovation, driving research and development in areas such as satellite technology, propulsion systems, and space-based infrastructure.

Economic impact is a central consideration in government initiatives, as space activities have the potential to generate significant returns and contribute to economic growth. Investments in space technology, satellite manufacturing, and launch services create jobs, stimulate the growth of the aerospace industry, and foster the development of a skilled workforce. The economic impact also extends to downstream industries, including telecommunications, agriculture, and environmental monitoring, which leverage space-based applications for enhanced efficiency and productivity.

Regulations emerge as a critical component in the governance of space activities, balancing the imperative for exploration with the need for responsible and sustainable practices. Governments formulate regulatory frameworks to manage space traffic, mitigate space debris, and ensure the responsible use of space resources. Regulatory bodies are tasked with overseeing commercial launches, satellite operations, and licensing procedures, creating a structured environment that fosters safety, security, and compliance with international norms.

Collaborative efforts on the global stage manifest in various forms, reflecting the interconnected nature of space exploration. International collaboration in space endeavors is not confined to geopolitical alliances but extends to a broader ethos of shared knowledge, resource pooling, and joint scientific pursuits. Collaborative projects, such as joint satellite development programs and international space missions, bring together the expertise and contributions of multiple nations, fostering a spirit of cooperation that transcends political boundaries.

Moreover, collaborative efforts are increasingly prevalent in commercial space activities. Public-private partnerships, where governments collaborate with private companies, facilitate the commercialization of space and drive economic growth. These partnerships leverage government investments to spur private innovation, creating a symbiotic relationship that benefits both sectors. The collaboration between government space agencies and private entities extends to launch services, satellite deployment, and the development of cutting-edge technologies.

The economic impact of government initiatives and collaborative efforts on the global stage is further amplified by the rise of the commercial space sector. Governments actively engage with private companies, encouraging innovation, providing incentives, and creating frameworks that foster a conducive environment for commercial space activities. The emergence of private space enterprises as key players in the global market contributes to economic growth, job creation, and the diversification of space-related business activities.

International collaboration in space extends beyond individual projects to collective endeavors that address global challenges. Initiatives like the Artemis program, led by NASA with international partnerships, exemplify collaborative efforts to explore the Moon and beyond. The pooling of resources, expertise, and infrastructure on a global scale not only advances scientific knowledge but also reinforces the idea that space exploration is a collective human endeavor with shared benefits for all.

In conclusion, the nexus of government initiatives, regulations, and collaborative efforts on the global stage shapes the economic impact and international collaboration in space exploration. As nations recognize the economic potential of space activities, governments play a crucial role in fostering an environment that stimulates innovation, ensures responsible practices, and facilitates collaboration among nations and private entities. The economic impact reverberates across industries, creating a ripple effect that extends beyond national borders, contributing to the shared aspirations of humanity to explore and harness the vast potential of outer space.

Future Prospects and Conclusion

Japan's Long-Term Goals in Space Exploration

Japan's long-term goals in space exploration weave a visionary spectrum that extends beyond the confines of our planet, outlining a trajectory marked by scientific discovery, technological innovation, and a steadfast commitment to advancing humanity's understanding of the cosmos. These goals, shaped by institutions such as the Japan Aerospace Exploration Agency (JAXA) and a vibrant network of public and private entities, encapsulate a multifaceted approach encompassing lunar exploration, Mars missions, space-based infrastructure, and international collaboration.

At the forefront of Japan's long-term goals is the ambition to establish a sustained presence on the Moon. The lunar exploration endeavors, exemplified by projects like the Smart Lander for Investigating Moon (SLIM) mission, signify a commitment to unlocking the mysteries of Earth's celestial neighbor. Japan envisions deploying robotic explorers and landers to conduct scientific investigations, studying the lunar surface, and paving the way for future human exploration. The Moon serves as both a scientific frontier and a stepping stone for deeper space exploration, aligning with the global aspiration to establish a sustainable human presence beyond Earth.

Mars missions stand as another pillar of Japan's long-term goals, reflecting the nation's commitment to unraveling the enigmas of the Red Planet. The Martian Moons Exploration (MMX) mission, spearheaded by JAXA, aims to explore the Martian moons Phobos and Deimos. This ambitious mission not only contributes to our understanding of Mars' geological evolution but also lays the groundwork for potential future human missions to the Martian system. Japan's endeavors in Mars exploration symbolize a foray into the outer reaches of our solar system, positioning the nation as a key player in the global quest for extraterrestrial knowledge.

Space-based infrastructure forms a crucial component of Japan's long-term goals, with an emphasis on the development of advanced satellites, space probes, and cutting-edge technologies. The Quasi-Zenith Satellite System (QZSS), a constellation of navigation satellites, exemplifies Japan's commitment to enhancing global positioning capabilities. Future prospects include the expansion of satellite constellations for various applications, from Earth observation to telecommunications, fostering economic growth and technological innovation.

International collaboration stands as a guiding principle in Japan's long-term goals, recognizing the collective nature of space exploration. Initiatives like the Lunar Gateway, a collaborative project involving NASA and international partners, exemplify Japan's role in contributing to a shared vision of human space exploration. By actively engaging in collaborative projects, Japan not only leverages global expertise but also positions itself as a key player in the collective efforts to explore and utilize space for the betterment of humanity.

Looking ahead, the future prospects of Japan's long-term goals in space exploration involve a continued commitment to innovation and adaptability. Technological advancements, including the development of next-generation launch vehicles and advancements in propulsion systems, are anticipated to propel Japan further into the forefront of space exploration. The nation's space industry is poised for continued growth, driven by collaborations with private enterprises and a focus on commercial space activities.

In conclusion, Japan's long-term goals in space exploration weave a narrative of scientific ambition, technological prowess, and international collaboration. As the nation charts a course toward the Moon, Mars, and beyond, it embraces a vision that extends beyond national boundaries, contributing to the collective human endeavor to explore the cosmos. The future prospects of Japan's space exploration endeavors hold the promise of groundbreaking discoveries, technological breakthroughs, and a continued commitment to the shared exploration of the vast and wondrous expanse of space.

Summarizing the Journey:
From Past Defeats to Current Triumphs in Satellite Innovation

The journey of Japan in satellite innovation encapsulates a narrative of resilience, technological evolution, and a transformative shift from past setbacks to current triumphs. It is a journey that has unfolded against the backdrop of historical challenges, marked by a commitment to learning, adapting, and positioning Japan as a leading player in the global arena of satellite technology. As we delve into the summary of this journey, we explore the pivotal moments, technological milestones, and future prospects that define Japan's trajectory in satellite innovation.

The narrative begins with an acknowledgment of past defeats, notably the setbacks experienced by Japan in its early attempts at space exploration. These challenges, including failed launches and setbacks in satellite programs, served as catalysts for introspection and strategic recalibration. Rather than succumbing to defeat, Japan leveraged these experiences as learning opportunities, laying the groundwork for a more resilient and forward-looking approach to satellite innovation.

The turning point in Japan's journey came with a renewed focus on technological advancement and international collaboration. The establishment of the Japan Aerospace Exploration Agency (JAXA) in 2003 marked a pivotal moment, consolidating the nation's space exploration efforts under a unified entity. This organizational shift facilitated a more streamlined and collaborative approach to satellite innovation, enabling Japan to pool its resources, expertise, and international partnerships for collective progress.

One of the cornerstones of Japan's satellite innovation is the H3 rocket developed by JAXA and Mitsubishi Heavy Industries. This advanced launch vehicle represents a triumph of engineering, incorporating modular design, enhanced payload capacity, and a commitment to cost-effectiveness. The successful launch of the H3 rocket not only demonstrated Japan's technological prowess but also positioned the nation as a reliable and competitive player in the global satellite launch market.

The narrative of Japan's journey in satellite innovation also intertwines with the broader dynamics of the space industry. The rise of affordable commercial vehicles, exemplified by SpaceX's reusable Falcon 9, has reshaped the landscape of satellite launches. In response, Japan strategically positioned itself to meet the demands of a rapidly evolving market, emphasizing adaptability, cost-effectiveness, and the development of next-generation launch vehicles.

The collaborative initiatives undertaken by Japan further amplify the triumphs in satellite innovation. Engaging in joint projects such as the LUPEX mission with India and contributions to the U.S.-led Artemis moon exploration program showcases Japan's role as a global collaborator in the pursuit of scientific knowledge and space exploration. These collaborations not only contribute to the collective understanding of the cosmos but also reinforce Japan's standing as a key partner in international space endeavors.

As we look towards future prospects, the trajectory of Japan's satellite innovation remains promising. The commitment to technological advancement is evident in plans for the development of next-generation satellites, advancements in propulsion systems, and the exploration of small satellite technologies. The goal of launching a lunar explorer for the joint Japan-India LUPEX project in 2025 and contributing to the U.S.-led Artemis program reflects a continued ambition to push the boundaries of space exploration.

In conclusion, the journey of Japan in satellite innovation is a testament to the nation's ability to transform past defeats into present triumphs. From historical challenges to the successful launches of the H3 rocket, Japan has demonstrated resilience, adaptability, and a forward-looking approach to space exploration. The collaborative spirit, technological advancements, and strategic positioning in the global space industry position Japan for continued success in satellite innovation. As the nation looks towards the future, the narrative unfolds with the promise of groundbreaking achievements, further contributions to international space endeavors, and a legacy of triumph in the ever-expanding cosmos.